# FESTUNG GUERNSEY

This book is unique amongst all volumes on German fortifications of the Second World War, in that it was written by the German forces themselves. It is the ultimate guide to their defensive planning and mindset.

The Channel Islands of Guernsey and Jersey nestle in the lee of the French coast. They are, however, British and in 1940 they became the first and only British territory to be occupied by the rapidly moving armies of Hitler's blitzkrieg.

Hitler swiftly became obsessed by his conquest. Determined that they should not be retaken he set in place a massive series of fortifications designed to make the islands impregnable fortresses or Festung, as part of the Atlantic Wall.

The Atlantic Wall ran for over three thousand miles, but by 1944 the tiny Channel Islands had consumed nearly 10% of all the concrete destined for the wall. They also contained the largest single garrison of the German Army and more heavy guns than 600 miles of Normandy coastline.

Field Marshall Rommel, in charge of the Normandy defences, pleaded with the Führer to release men and materials from the islands to Normandy. Hitler in a fury forbade him to raise the matter again and senior Reich officers began to talk of Hitler's "inselwahn" his "island madness".

In 1944 Lieutenant Colonel Hubner was charged with making a record of the immense fortifications. His team was drawn from the Divisionskartenstelle, the Divisional Cartographic Section, with some fourteen non-commissioned officers working across: drawing, photography, cartography, calligraphy and printing. The result is a stunning and comprehensive picture of the fortifications and a complete guide to their workings.

Festung Guernsey consists of 22 chapters and was originally published as a limited edition of 135, two-volume sets, bound in leather. The original work, being made by hand was only printed on the right hand pages, this means we have been able to provide a full translation on the left hand page, while retaining the original layout.

This paperback version will consist of 10 separate volumes, each consisting of 1,2 or 3 chapters and replicates the page numbering of the original edition.

Volumes are being published every four months with the 10th being completed in May 2015 on the 70th anniversary of the liberation of the islands by Force 135.

By 1944 Guernsey was the most fortified place in the world. These immense fortifications were built using slave labour. Please pause for a moment before turning the page and remember the men of many nationalities upon whose privations, ill treatment and lives this Festung was built.

# The Festung Guernsey part work edition, part by part.

## PART 1

**Covers the coastal fortifications along the east coast
from St Martins Point to St Sampsons.**

## PART 2

Covers the coastal fortifications along the east and north coasts
from St Samspsons to Grande Havre.

## PART 3

Covers the coastal fortifications along the west coast
from Grande Havre to Fort Sausmarez.

## PART 4

Covers the coastal fortifications along the west and south coasts
from Rocquaine to Corbiere.

## PART 5

The final coastal section covers the fortifications along the south
and east coasts from Corbiere to Fort George.

## PART 6

The English Garrison on Guernsey, a history of the island's fortifications.

## PART 7

General Information about Guernsey, a history of the island,
its laws and customs.

## PART 8

Tactical Review of the Fortified areas & Fortified Structures,
detailing the garrison, and the bunker and emplacement designs.

## PART 9

Weapons deployed & Mirus Battery, list and photographs of all the weaponry
and a chapter on the immense Mirus battery.

## PART 10

Deployment of Artillery & Anti-Aircraft Artillery, comprises details and maps
of every artillery and anti-aircraft battery.

# FESTUNG GUERNSEY

CHAPTERS: 3.1 & 3.2

# ST PETER PORT HARBOUR

## from Les Terres Point to Salerie Corner

HAFEN
ST PETER PORT
von Terres Point bis Peterseck

St Peter Port Harbour

from Les Terres Point to Salerie Corner

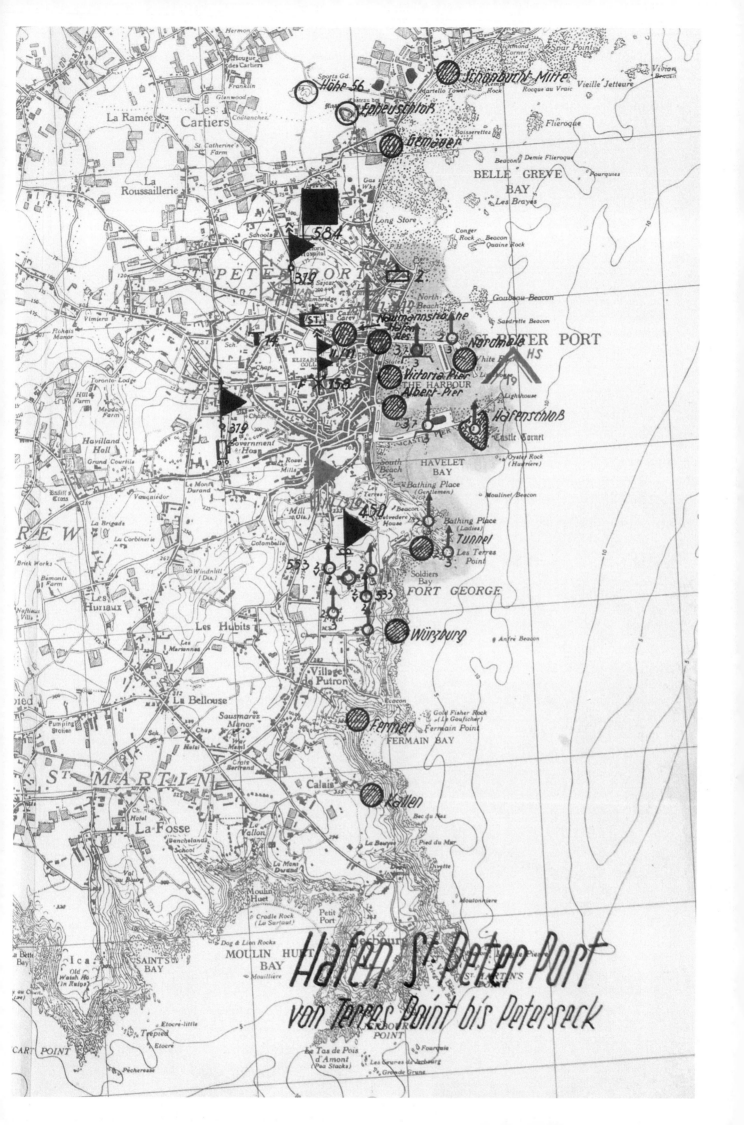

Hafen St. Peter Port
von Terres Point bis Peterseck

# St Peter Port Harbour

1:5000

## Legend

⊥ 10.5 cm gun in casemate
↑ 10.5 cm gun in field position
🮤 7.5 cm gun in embrasure
⊥ 4.7 cm anti-tank gun in casemate
⊤ 3.7 cm and 5 cm anti-tank gun in field position
⊥ 3.7 cm tank gun
⤊ 3.7 cm tank gun and machine-gun
⊥ Armoured vehicle with machine-gun
∠ 8 cm and 5 cm mortars
◢ M19 (Maschinengranatwerfer, a fully-automatic 5 cm mortar)
◎ Tobruk pit
ᒫ Spigot mortar
✿ Multi-loopholed turret
ⴹ Searchlights
⸬ Mines

**Firing zones**:

◁ 10.5 cm gun
◁ 4.7 cm anti-tank gun
⬭ Anti-aircraft gun
⬭ Mortar and M19
⬭ Machine-gun
⬭ Spigot mortar

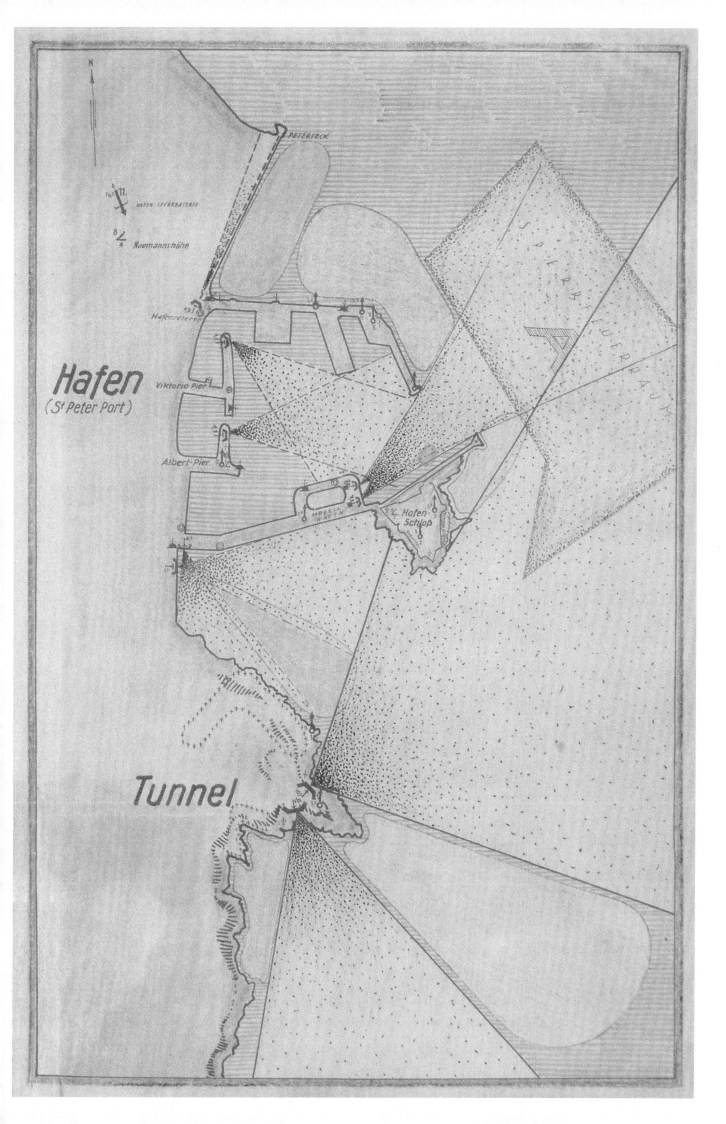

# TUNNEL Resistance Nest

**1.)** **Deployment** at the Harbour Company's right-hand boundary (sector boundary between infantry regiments 584 and 583) between Fort George and Havelet Bay.

**2.)** **Contingent**:

| | | |
|---|---|---|
| | 1 NCO | 14 men |
| naval company (only at stage two alert) | 1 petty officer | 11 ratings |
| anti-aircraft artillery | 6 NCOs | 31 men |

**3.)** **Weapons**:

own:
- 2 10.5 cm casemate guns
- 2 light machine guns on bunker mount
- 1 60 cm searchlight (may be requisitioned from the anti-aircraft artillery to be deployed against airborne targets)
- 1 defensive flame-thrower 42

VGAD:
- 1 light machine gun (Dutch)

anti-aircraft artillery:
- 3 3.7 cm anti-aircraft guns
- 3 2 cm anti-aircraft guns
- 2 machine guns MG15
- 1 machine gun MG34
- 1 60 cm searchlight (being installed)

**4.)** **Military objectives**: The crew of the resistance nest will defend itself, to the last man, against enemy troops attacking from sea, air or land, under the direction of the infantry base's officer in charge.

Specifically, enemy troops landing in the <u>bays on both sides of Les Terres Point</u> are to be resisted using all available weapons.

Defence actions must focus on possible landing places below Fort George and in Havelet Bay.

**5.)** **Operations**: Enemy troops landing in Soldiers Bay are to be engaged using one 10.5 cm casemate gun and the machine gun MG34 from opposing positions. The mortar battery at Cambridge Park will cover the dead space. The crew at Fort George will provide additional defence by activating roll bombs. The northern casemate gun is to fire on enemy craft in the Little Russell; in addition, enemy troops landing in Havelet Bay will be fought using the machine gun MG34.

The defence will be supported by a 3.7 cm anti-aircraft platoon located at Les Terres Point, a 2 cm anti-aircraft platoon stationed at the Bathing Place (Ladies), the Havelet Bay resistance nest and the Castle Cornet strong point. By night, one 60 cm searchlight belonging to the infantry and one belonging to the anti-aircraft artillery at Les Terres Point will be used.

# Widerstandsnest T U N N E L

1.) **Einsatz** an rechter Grenze Hafenkompanie (Abschnittsgrenze zwischen Gren.Rgt. 584 und 583) zwischen Georgsfeste und Havelet Bay.

2.) **Stärke:**                                   1 Uffz.   14 Mannschaften
     Marine Komp.
     (nur bei Alarmstufe II)                 1 Uffz.   11 Mannschaften
     Flak                                          6 Uffz.   31 Mannschaften

3.) **Waffen:**
     eigene:   2  1o,5 cm Kas.Kan.
                   2  le.M.G. auf Bunkerlafette
                   1  6o cm Scheinwerfer (kann von der Flak gegen
                                                    Luftziele angefordert werden)
                   1  Abwehrflammenwerfer 42
     VGAD:    1  le.M.G. (holl.)
     Flak:      3  3,7 cm Flak
                   3  2 cm Flak
                   2  M.G.15
                   1  M.G.34
                   1  6o cm Scheinwerfer (im Einbau).

4.) **Kampfauftrag:** Die Besatzung verteidigt sich gegen von See, aus der Luft und von Land angreifenden Feind in ihrem Widerstandsnest bis zum letzten Mann unter Führung des Infanterie-Stützpunktführers.

Im besonderen ist die Anlandung von Feindkräften in den **Buchten beiderseits Felsenspitze** (Les Terres Point) mit allen Waffen zu bekämpfen.

Verteidigungsschwerpunkt: Landestellen unterhalb Georgs= feste und in der Havelet Bay.

5.) **Kampfführung:** In der Soldaten-Bucht (Soldiers Bay) lan= dender Gegner ist mit einer 1o,5 cm Kas.Kan. und aus verbreiteter Stellung mit M.G.34 zu bekämpfen. In den toten Raum wirkt das Feuer der Gran.Werf.-Batterie Nau= mannshöhe. Zusätzliche Abwehr durch Auslösen der Roll= bomben durch Besatzung Georgsfeste. Feindboote im kleinen Rüssel sind mit der nördlichen Kas.Kan., in der Havelet Bay landender Gegner darüberhinaus mit M.G.34 zu bekämp= fen.
Unterstützt wird der Abwehrkampf durch einen 3,7 cm Flak-Zug auf Felsenspitze (Les Terres Point) und einen 2 cm Flak-Zug beim Bathing Place (Ladies), durch das Wider= standsnest Havelet Bay und den Stützpunkt Hafenschloß (Cornet Castle), bei Nacht durch je einen 6o cm Schein= werfer der Inf. und der Flak auf Felsenspitze (Les Terres Point).

# HAVELET BAY Resistance Nest

**1.)** **Deployment** between Tunnel and Model Yacht Pond resistance nests.

**2.)** **Contingent**:                   1 NCO          7 men

naval company

(only at stage two alert)              2 petty officers     5 ratings

**3.)** **Weapons**:

own:                     1 machine gun MG311 (f) in revolving turret

1 machine gun on bunker mount

2 4.7 cm anti-tank guns (t) with machine guns (t)

(being installed)

1 handheld searchlight

navy:                    1 machine gun MG34

**4.)** **Military objectives**: The crew of the resistance nest will defend itself, to the last man, against enemy troops attacking from sea, air or land.

Specifically, enemy troops landing at <u>Havelet Bay</u> and at the <u>entrance to the South Mole</u> are to be resisted using all available weapons.

Defence actions must focus on possible landing places in Havelet Bay.

**5.)** **Operations**: Enemy troops landing in Havelet Bay are to be engaged using all weapons (one machine gun MG34, one machine gun MG311 (f) in revolving turret). Action is to be taken against landing craft and tanks attempting to land, using, in particular, both 4.7 cm anti-tank guns (t) with machine guns (t) (being installed). The defence will be supported by a 10.5 cm casemate gun and a machine gun MG34 located at the resistance nest at Tunnel, two machine guns MG34 and two 5 cm mortars (f) located at Castle Cornet, and one each of the 3.7 cm and 2 cm anti-aircraft guns from the set stationed at the Tunnel resistance nest and Castle Cornet strong point.

Due to the high coastal walls, the beach is secure against mechanised attack. Tanks approaching from the land side will be fought with anti-tank weapons.

## Widerstandsnest  H A V E L E T - B A Y

1.) <u>Einsatz</u> zwischen Widerstandsnest Tunnel und Widerstands=
nest Modellhafen.

2.) <u>Stärke:</u>                                    1 Uffz.  7 Mannschaften
Marine Komp.
(nur bei Alarmstufe II)      2 Uffz.  5 Mannschaften

3.) <u>Waffen:</u>
Eigene:     1  M.G. 311 (f) im Drehturm
1  M.G. auf Bunkerlafette
2  4,7 cm Pak (t) mit M.G. (t) (im Einbau)
1  Handscheinwerfer
Marine:     1  M.G.34

4.) <u>Kampfauftrag:</u> Die Besatzung verteidigt sich gegen von
See, aus der Luft und von Land angreifenden Feind in
ihrem Widerstandsnest bis zum letzten Mann.

Im besonderen ist die Anlandung von Feindkräften in der
<u>Havelet Bay</u> und dem <u>Eingang zur Südmole</u> mit allen Waffen
zu bekämpfen.

Verteidigungsschwerpunkt: Landestellen in der Havelet
Bay.

5.) <u>Kampfführung:</u> In der Havelet Bay landender Gegner ist
mit allen Waffen (1 M.G.34, 1 M.G.311 (f) im Drehturm)
zu bekämpfen, Einsatz gegen Landeboote und landende
Panzer insbesondere mit den beiden 4,7 cm Pak (t) mit
M.G. (t) (im Einbau).  Unterstützt wird der Abwehrkampf
durch eine 1o,5 cm Kas.Kan. und ein M.G.34 vom Wider=
standsnest Tunnel, 2 M.G.34 und 2  5cm Gran.Werf.(f)
vom Hafenschloß (Cornet Castle) und je einen 3,7 cm und
2 cm Flakzug vom Widerstandsnest Tunnel und Stützpunkt
Hafenschloß.

Der Strand ist durch die hohen Ufermauern panzersicher,
Panzer von Land her werden mit Panzernahkampfmitteln
bekämpft.

# MODEL YACHT POND Resistance Nest

**1.)** **Deployment** between Havelet Bay resistance nest and Castle Cornet strong point.

**2.)** **Contingent**:

|  |  |  |
|---|---|---|
|  | 1 NCO | 8 men |
| naval company | | |
| (only at stage two alert) | 2 petty officers | 10 ratings |
| anti-aircraft artillery | 1 NCO | 5 men |

**3.)** **Weapons**:

| own: | 2 7.5 cm guns (p) |
|---|---|
|  | 1 machine gun MG34 |
| navy: | 1 machine gun MG34 |
| anti-aircraft artillery: | 1 3.7 cm anti-aircraft gun |

**4.)** **Military objectives**: The crew of the resistance nest will defend itself, to the last man, against enemy troops attacking from sea, air or land.

Specifically, enemy vessels approaching the <u>harbour entrance</u> and attacks from the land side are to be resisted using all available weapons.

Defence actions must focus on the entrance to the harbour.

**5.)** **Operations**: Enemy vessels breaching the harbour entrance are to be engaged using both 10.5 cm casemate guns. In addition, the machine gun MG34 and the machine gun MG311 (f) will be deployed against any enemy forces invading the harbour basin. In particular, the resistance nest is charged with the protection of the Castle Cornet strong point to the land side. When used to combat enemy troops approaching from the land side, the 7.5 cm guns are to be transferred from their embrasures to their designated open positions. Anti-tank weapons, to engage approaching tanks, and an assault detachment to fight in the intermediate terrain, are to be held in readiness.

The defence will be supported by two machine guns MG34, one machine gun MG15 and two 5 cm mortars (f) located at Castle Cornet, by the resistance nests at Havelet Bay, Albert Pier, Victoria Pier, North Mole, and also by anti-aircraft artillery deployed at Castle Cornet and at North Mole. The anti-aircraft guns are included in the barrage firing plan.

# Widerstandsnest MODELLHAFEN

1.) <u>Einsatz</u> zwischen Widerstandsnest Havelet Bay und
    Stützpunkt Hafenschloß.

2.) <u>Stärke:</u>                                    1 Uffz.    8 Mannschaften
    Marine Komp.
    (nur bei Alarmstufe II)             2 Uffz.   10 Mannschaften
    Flak                                            1 Uffz.    5 Mannschaften

3.) <u>Waffen:</u>
    Eigene:   2  7,5 cm Kan. (p)
              1  M.G.34

    Marine:   1  M.G.34

    Flak:     1  3,7 cm Flak.

4.) <u>Kampfauftrag:</u> Die Besatzung verteidigt sich gegen von
    See, aus der Luft und von Land angreifenden Feind in
    ihrem Widerstandsnest bis zum letzten Mann.

    Im besonderen ist das Eindringen von feindlichen Fahr=
    zeugen in die <u>Hafeneinfahrt</u> und ein Angriff von Land her
    mit allen Waffen zu bekämpfen.

    Verteidigungsschwerpunkt: Hafeneinfahrt.

5.) <u>Kampfführung:</u> In die Hafeneinfahrt eindringende Feind-
    fahrzeuge sind mit den beiden 7,5 cm Kanonen zu bekäm=
    fen, darüberhinaus Einsatz des M.G.34 und des M.G.311(f)
    gegen jeden im Hafenbecken auftauchenden Feind. Insbe=
    sondere übernimmt das Widerstandsnest den Schutz der
    Landseite des Stützpunktes Hafenschloß. Beim Einsatz
    gegen Feind von Land sind die 7,5 cm Kanonen aus ihren
    Schartenständen in die vorbereiteten offenen Stellungen
    herauszuziehen. Panzernahkampfmittel gegen auftauchende
    Panzer und ein Stoßtrupp zum Kampf im Zwischengelände
    sind bereitzuhalten.

    Unterstützt wird der Kampf durch 2 M.G.34, 1 M.G.15 und
    2 5 cm Gr.W. (f) von Hafenschloß, durch die Widerstands=
    nester Havelet Bay, Albert Pier, Viktoria Pier, Nordmole,
    sowie durch die auf Hafenschloß und auf der Nordmole
    eingesetzte Flak. Flak ist in den Sperrfeuerplan einbe=
    zogen

# CASTLE CORNET Strong Point

**1.)** <u>**Deployment**</u> between Havelet Bay and the harbour entrance.

**2.)** <u>**Contingent**</u>:

| | | |
|---|---|---|
| own (armourers) | 1 NCO | 2 men |
| navy (only at stage two alert) | 2 petty officers | 5 ratings |
| navy (miscellaneous) | | 5 ratings |
| anti-aircraft artillery | 1 officer | 14 NCOs | 36 men |

**3.)** <u>**Weapons**</u>:

own:
2 5 cm mortars (f)
2 handheld searchlights
15 roll bombs
3 emplaced grenades
1 radio-telephone (C)

naval company:
2 machine guns MG34 on bunker mount

naval (miscellaneous):
1 machine gun (p)

anti-aircraft artillery:
2 3.7 cm anti-aircraft guns
3 2 cm anti-aircraft guns
1 60 cm searchlight

**4.)** <u>**Military objectives**</u>: The crew of the resistance nest will defend itself, to the last man, against enemy troops attacking from sea, air or land, under the direction of Castle Cornet's commanding officer (anti-aircraft battery commander).

Specifically, enemy troops landing at <u>Havelet Bay</u>, at South Mole and at the <u>entrance to the harbour</u> are to be resisted using all available weapons.

Defence actions must focus on Havelet Bay and the entrance to the harbour.

**5.)** <u>**Operations**</u>: The strong point will defend itself on all sides against landing enemy troops. Landing craft are to be combated using anti-aircraft weapons (2 cm and 3.7 cm). In addition, landing enemy forces will be fought with two machine guns MG34, one machine gun MG15, two 5 cm mortars (f) and roll bombs emplaced all around. In case of attack from the land side, the bridge is to be blown up by activating the emplaced grenades.

Assault detachments are to be deployed to engage any enemy forces which have broken through and any enemy troops on adjacent terrain at Model Yacht Pond.

Using its fire power, the base will support the defence of the resistance nests at Tunnel, Havelet Bay and Model Yacht Pond, as well as covering the entire harbour basin.

## Stützpunkt H A F E N S C H L O S S (Cornet Castle)

1.) <u>Einsatz</u> zwischen der Havelet Bay und der Hafeneinfahrt.

2.) <u>Stärke:</u>
| | | |
|---|---|---|
| eigene (Waffenwarte) | 1 Uffz. | 2 Mannschaften |
| Marine (nur bei Alarmstufe II) | 2 Uffz. | 5 Mannschaften |
| Marine (sonstige) | | 5 Mannschaften |
| Flak | 1 Offz. 14 Uffz. | 36 Mannschaften |

3.) <u>Waffen:</u>
| | | |
|---|---|---|
| eigene: | 2 | 5 cm Gr.W. (f) |
| | 2 | Handscheinwerfer |
| | 15 | Rollbomben |
| | 3 | Granaten eingebaut |
| | 1 | Funksprechgerät (C) |
| Marine Komp. | 2 | M.G.34 auf Bunkerlafette |
| Marine (sonstige) | 1 | M.G. (p) |
| Flak | 2 | 3,7 cm Flak |
| | 3 | 2 cm Flak |
| | 1 | Scheinwerfer 60 cm. |

4.) <u>Kampfauftrag:</u> Die Besatzung verteidigt sich unter der Führung des Kommandanten des Hafenschlosses (Batterieführer Flak) gegen von See, Land und aus der Luft angreifenden Feind in ihrem Stützpunkt bis zum letzten Mann.

Im besonderen ist die Anlandung von Feindkräften in der <u>Havelet Bay</u>, an der Südmole und in der <u>Hafeneinfahrt</u> mit allen Waffen zu bekämpfen.

Verteidigungsschwerpunkt: Havelet Bay und Hafeneinfahrt.

5.) <u>Kampfführung:</u> Der Stützpunkt verteidigt sich nach allen Seiten gegen anlandenden Feind. Bekämpfung der Landeboote durch die Flakwaffen (2 cm und 3,7 cm), Bekämpfung des landenden Gegners darüberhinaus mit 2 M.G.34, 1 M.G.15, 2 5 cm Gr.W. (f) und rundum eingebaute Rollbomben. Bei Angriff von Land ist die Brücke durch Auslösen der eingebauten Granaten zu sprengen.

Gegen eingedrungenen Feind oder Feind im Nachbargelände (Modellhafen) sind Stoßtrupps einzusetzen.

Der Stützpunkt greift in den Abwehrkampf der Widerstandsnester Tunnel, Havelet-Bay und Modellhafen und im gesamten Hafenbecken mit seinen Waffen unterstützend ein.

# ALBERT PIER Resistance Nest

**1.)** **Deployment** between Havelet Bay and Victoria Pier.

**2.)** **Contingent**:

| | | |
|---|---|---|
| own (armourers): | 1 NCO | 2 men |
| naval company | | |
| (only at stage two alert): | 4 petty officers | 10 ratings |
| navy (miscellaneous): | 1 petty officer | 3 ratings |
| anti-aircraft artillery: | 2 NCOs | 2 men |

**3.)** **Weapons**:

| | |
|---|---|
| naval company: | 1 4.7 cm anti-tank gun (t) with machine gun (t) |
| | 1 5 cm mortar |
| | 2 machine guns MG34 (one on bunker mount) |
| | 1 machine gun MG311 (f) in revolving turret |
| | 1 medium-range flame-thrower |
| navy (miscellaneous): | 1 2 cm anti-aircraft gun |
| anti-aircraft artillery: | 1 60 cm searchlight |
| | 1 machine gun MG15 |

**4.)** **Military objectives**: The crew of the resistance nest will defend itself, to the last man, against enemy troops attacking from sea, air or land.

Specifically, enemy vessels breaching the <u>harbour entrance</u> and enemy approach from the land side are to be resisted using all available weapons.

Defence actions must focus on the entrance to the harbour.

**5.)** **Operations**: The resistance nest will combat enemy forces entering the harbour using all available weapons. The 4.7 cm anti-tank gun will mainly cover enemy landing craft. The two machine guns MG34, the machine gun MG311 (f) and the 5 cm mortar will be used to support the fight against enemy forces landing on the jetties (moles) and piers. The medium flame-thrower is designated predominantly for defence of the land side. In the event of a tank attack, anti-tank weapons are to be used. Should enemy troops have reached the intermediate terrain, an assault detachment will be deployed.

The defence will be supported by all resistance nests located in the vicinity of the harbour. Low-level attacks by enemy aircraft are to be repulsed using the 2 cm anti-aircraft gun and the anti-aircraft artillery stationed at the harbour.

# Widerstandsnest ALBERT PIER

1.) <u>Einsatz</u> zwischen Havelet Bay und Viktoria Pier.

2.) <u>Stärke:</u>

| | | |
|---|---|---|
| eigene (Waffenwarte) | 1 Uffz. | 2 Mannschaften |
| Marine Komp.(nur bei Alarmstufe II) | 4 Uffz. | 10 Mannschaften |
| Marine (sonstige) | 1 Uffz. | 3 Mannschaften |
| Flak | 2 Uffz. | 2 Mannschaften |

3.) <u>Waffen:</u>

Marine Komp.:
  1  4,7 cm Pak (t) mit M.G. (t)
  1  5 cm Gr.W.
  2  M.G. 34 (davon 1 auf Bunkerlafette)
  1  M.G. 311(f) im Drehturm
  1  mittl.Flammenwerfer

Marine (sonstg): 1 2cm Flak

Flak
  1  6o cm Scheinwerfer
  1  M.G. 15

4.) <u>Kampfauftrag:</u> Die Besatzung verteidigt sich gegen von See, Land und aus der Luft angreifenden Feind in ihrem Widerstandsnest bis zum letzten Mann.

Im besonderen ist die Einfahrt von feindlichen Fahrzeugen in die <u>Hafeneinfahrt</u> und das Eindringen von Landseite mit allen Waffen zu bekämpfen.

Verteidigungsschwerpunkt: Hafeneinfahrt.

5.) <u>Kampfführung:</u> Das Widerstandsnest bekämpft mit allen Waffen durch die Hafeneinfahrt eindringenden Gegner, Landeboote vor allem mit der 4,7 cm Pak. Der Abwehrkampf gegen auf den Piers und Molen gelandeten Feind ist mit den beiden M.G.34, dem M.G.311 (f) und dem 5 cm Gr.W. zu unterstützen. Für die Abwehr nach Landseite ist besonders der mittl. Flammenwerfer vorgesehen. Bei Panzerangriff Einsatz von Panzernahkampfmitteln, bei Feind im Zwischen= gelände Einsatz eines Stoßtrupps.

Der Abwehrkampf wird durch sämtliche im Hafen eingesetz= ten Widerstandsnester unterstützt, Tieffliegerangriffe werden durch das 2 cm Flakgeschütz und durch die im Hafen eingesetzte Flak abgewehrt.

# VICTORIA PIER Resistance Nest

**1.)** <u>**Deployment**</u> between Old Harbour and Albert Pier.

**2.)** <u>**Contingent**</u>:

| | | |
|---|---|---|
| own (armourers): | 2 men | |
| naval company | | |
| (only at stage two alert): | 1 petty officer | 12 ratings |

**3.)** <u>**Weapons**</u>:

naval company:
- 1 4.7 cm anti-tank gun (t) with machine gun (t)
- 1 machine gun MG311 (f) in revolving turret
- 1 machine gun MG34
- 1 medium-range flame-thrower

**4.)** <u>**Military objectives**</u>: The crew of the resistance nest will defend itself, to the last man, against enemy troops attacking from sea, air or land.

Specifically, enemy vessels penetrating the <u>harbour entrance</u> and enemy approach from the land side are to be resisted using all available weapons.

Defence actions must focus on the entrance to the harbour.

**5.)** <u>**Operations**</u>: The resistance nest will combat enemy forces entering the harbour using all available weapons. The machine guns will be used to support the fight against enemy forces landing on the jetties (moles) and piers. Any attack on this resistance nest is to be repulsed by using portable infantry weapons; the medium-range flame-thrower is to be used, above all else, in the event of an attack from the land side. Should there be fighting in the intermediate terrain, the resistance nest will intervene, deploying an assault detachment.

The defence will be supported by all resistance nests located in the vicinity of the harbour, in particular by those at Albert Pier and Old Harbour.

# Widerstandsnest V I K T O R I A   P I E R

1.) <u>Einsatz</u> zwischen Hafenreserve und Albert Pier.

2.) <u>Stärke:</u>

    eigene (Waffenwarte)                        2 Mannschaften

    Marine Kp.(nur bei Alarmstufe II)

                                   1 Uffz. 12 Mannschaften

3.) <u>Waffen:</u>

    Marine Komp. :   **1** Pak 4,7 cm (t) mit M.G.(t)
                     **1** M.G.311 (f) im Drehturm
                     **1** M.G.34
                     **1** mittl. Flammenwerfer

4.) <u>Kampfauftrag:</u> Die Besatzung verteidigt sich gegen von See, Land und aus der Luft angreifenden Feind in ihrem Widerstandsnest bis zum letzten Mann.

Im besonderen ist die Einfahrt von feindlichen Fahrzeu= gen in die <u>Hafeneinfahrt</u> und das Eindringen von Land= seite mit allen Waffen zu bekämpfen.

Verteidigungsschwerpunkt: Hafeneinfahrt.

5.) <u>Kampfführung:</u> Das Widerstandsnest bekämpft mit allen Waffen durch die Hafeneinfahrt eindringenden Gegner. Der Abwehrkampf gegen auf den Piers und Molen gelande= ten Feind ist mit den M.G. zu unterstützen. Angriff auf das eigene Widerstandsnest ist durch die beweglich ein= zusetzenden Inf.Waffen abzuwehren, der mittl. Flammen= werfer vor allem gegen Landangriff einzusetzen. Bei Kampf im Zwischengelände greift das Widerstandsnest mit einem Stoßtrupp ein.

Unterstützt wird der Abwehrkampf durch sämtl. im Hafen eingesetzte Widerstandsnester, vor allem durch Albert- Pier und Hafenreserve.

# NORTH MOLE Resistance Nest

**1.)** __Deployment__ between the harbour entrance and Old Harbour.

**2.)** __Contingent__:

| | | |
|---|---|---|
| own (armourers): | | 2 men |
| navy (only at stage two alert): | 1 petty officer | 7 ratings |
| navy (miscellaneous): | variable, a minimum of 10 in number | |
| anti-aircraft artillery: | 1 officer | 8 NCOs | 32 men |

**3.)** __Weapons__:

own:
2 machine guns MG311 (f) in revolving turret
2 medium-range flame-throwers
1 loophole searchlight

naval company:
2 machine guns MG34

anti-aircraft artillery:
3 3.7 cm anti-aircraft guns
3 2 cm anti-aircraft guns
2 60 cm searchlights

**4.)** __Military objectives__: The crew of the resistance nest will defend itself, to the last man, against enemy troops attacking from sea, air or land, under the direction of the North Mole's commanding officer (the most senior member of the anti-aircraft platoon).

Specifically, enemy troops landing between the <u>entrance to the harbour and Salerie Corner</u> are to be resisted using all available weapons.

Defence actions must focus on the entrance to the harbour.

**5.)** __Operations__: The resistance nest will use all its weapons to repel any enemy landing in the harbour. In particular, a machine gun MG34 is to fire onto the harbour entrance from the bunker at the head of the mole; a machine gun MG311 (f) in a revolving turret at the head of the mole is to cover the outer wall of the North Mole; the second machine gun MG34 and the revolving turret in the dog-leg of the North Mole are to deployed against enemy troops docking on the outside perimeter of the mole.

In addition, landed enemy forces are to be engaged in close combat using two medium flame-throwers. An assault detachment is to be kept in readiness to fight in the intermediate terrain at Middle Pier and Slaughterhouse Pier. Defence of neighbouring positions in the harbour and along the coast road is to be supported by using the fire power of all available weapons. The anti-aircraft artillery will help combat enemy landing craft and landed enemy forces using assault detachments, provided their main task of securing the airspace above the harbour permits. Supporting fire will be provided by all resistance nests located in the harbour, especially by Castle Cornet and Old Harbour.

# Widerstandsnest N O R D M O L E

1.) <u>Einsatz</u> zwischen Hafeneinfahrt und Hafenreserve.

2.) <u>Stärke</u>:

      eigene (Waffenwarte):              2 Mannschaften

      Marine (nur bei Alarmstufe II):1 Uffz.7 Mannschaften

      Marine (sonstige):   wechselnd,mindestens 10 Köpfe

      Flak:             1 Offz.  8 Uffz.32Mannschaften

3.) <u>Waffen</u>:

      eigene:        2  M.G.311 (f) Drehturm
                   2  m.Flammenwerfer
                   1  Schartenscheinwerfer

      Marine-Komp.: 2  M.G.34

      Flak:         3  3,7 cm Flak
                   3  2 cm Flak
                   2  6o cm Scheinwerfer

4.) <u>Kampfauftrag</u>: Die Besatzung verteidigt sich unter Füh=
rung des Kommandanten der Nordmole (ältester Flakzug=
führer) gegen von See, Land und aus der Luft angreifen=
den Feind in ihrem Widerstandsnest bis zum letzten Mann.

Im besonderen ist die Anlandung von Feindkräften zwi=
schen <u>Hafeneinfahrt und Peterseck</u> mit allen Waffen zu
bekämpfen.

Verteidigungsschwerpunkt: Hafeneinfahrt.

5.) <u>Kampfführung</u>: Das Widerstandsnest greift mit sämtlichen
Waffen in die Abwehr gegen einen im Hafen anlandenden
Gegner ein. Insbesondere wirkt es mit einem M.G.34 vom
Bunker Molenkopf in die Hafeneinfahrt, mit dem M.G.311
(f) im Drehturm am Molenkopf längs der Außenmauer der
Nordmole, mit dem zweiten M.G.34 und dem Drehturm am
Nordmolenknie gegen an der Außenseite der Mole anlegen=
den Feind.

Gelandeter Feind ist im Nahkampf darüberhinaus mit 2
mittlere Flammenwerfer zu bekämpfen. Zum Kampf im
Zwischengelände (Mittelpier, Schlachthofpier) ist ein
Stoßtrupp bereitzuhalten. Abwehrkampf der Nachbarstütz=
punkte im Hafen und auf der Uferstrasse ist mit Feuer
aller Waffen zu unterstützen. Die den Luftraum sichern=
de Flak greift in die Bekämpfung der Landeboote und ge=
gen gelandeten Feind mit Stoßtrupps ein, sofern die
Luftlage es zuläßt. Unterstützung des Feuerkampfes er=
folgt durch sämtliche im Hafen liegende Widerstandsne=
ster, insbesondere durch Hafenschloß und Hafenreserve.

# CAMBRIDGE PARK Resistance Nest

## (Mortar battery)

**1.)** Deployment in the rear of the company sector (at the eastern edge of Cambridge Park) with a radius of action covering the entire company area.

**2.)** <u>**Contingent**</u>:

|  |  |  |
|---|---|---|
| own: | 1 NCO | 5 men |
| Headquarters Company | 1 NCO | 5 men |

**3.)** <u>**Weapons**</u>:

own:
3 8 cm mortars
1 light machine gun MG34

**4.)** <u>**Military objectives**</u>: The crew of the resistance nest will defend itself, to the last man, against enemy troops attacking from sea, air or land.

Specifically, it will provide fire cover for the coastal stretch between Soldiers Bay and Belle Grève Bay.

Defence actions must focus on the <u>entrances to the harbour and landing places</u>.

**5.)** <u>**Operations**</u>: Enemy troops landing on the stretch between Soldiers Bay and Belle Grève Bay are to be engaged using the 8 cm mortars, depending on the situation.

In the event of enemy forces penetrating into the intermediate terrain, part of the crew is to be deployed as an assault detachment to fight them. The machine gun MG34 is to be deployed for the defence of the periphery. Should the whole of Cambridge Park need to be defended, the crew are subject to the command of 11th Battery, Army Coastal Artillery Regiment 1265.

Should the harbour company be required in other areas, one 8 cm mortar and one machine gun MG34 will remain available. A new crew will be provided by the Headquarters Company, Infantry Regiment 584.

# Widerstandsnest N A U M A N N S H Ö H E
## (Granatwerfer-Bttr.)

1.) Einsatz im rückwärtigen Kompanie-Abschnitt (Ostrand Naumannshöhe) mit Wirkungsbereich im gesamten Kom= panie - Bereich.

2.) <u>Stärke:</u>

| | | |
|---|---|---|
| eigene: | 1 Uffz. | 5 Mannschaften |
| Stabskompanie | 1 Uffz. | 5 Mannschaften |

3.) <u>Waffen:</u>

eigene: 3 8 cm Gr.Werf.

1 le.M.G.34

4.) <u>Kampfauftrag:</u> Die Besatzung verteidigt sich gegen von See, Land und aus der Luft angreifenden Feind in ihrem Widerstandsnest bis zum letzten Mann.

Im besonderen wirkt sie mit Feuer auf den Küstensaum von der Soldatenbucht (Soldiers Bay) bis zur Schönbucht (Belle Greve Bay).

Verteidigungsschwerpunkt: <u>Einfahrten und Landestellen.</u>

5.) <u>Kampfführung:</u> Anlandender Feind in dem Geländestreifen von Soldatenbucht bis Schönbucht ist je nach Lage durch Feuer der 8 cm Gr.Werf. zu bekämpfen.

Bei Feindeinbruch in das Zwischengelände sind Teile der Besatzung als Stoßtrupp gegen diesen Gegner einzu= setzen. Zur Rundumverteidigung wird das M.G.34 eingesetzt. Bei Gesamtverteidigung der Naumannshöhe ist die Mannschaft der 11./H.K.A.R.1265 unterstellt.

Bei Herauslösung der Hafenkompanie verbleiben 1 8 cm Gr.Werfer und 1 M.G.34. Neue Besatzung durch Stabskom= panie Gren.Rgt. 584.

# OLD HARBOUR Resistance Nest

## (also Company Command Post)

**1.)** **Deployment** at the harbour approach, including assault detachments throughout the entire company sector.

**2.)** **Contingent**:

| | | | |
|---|---|---|---|
| own: | 2 officers | 7 NCOs | 25 men |
| navy (only at stage two alert): | (2) naval officers | 3 petty officers | 9 ratings |

In the case of combat this contingent will be joined by available
forces from the VGAD and the harbour command, and from the vessels anchored in harbour.

**3.)** **Weapons**:

| | |
|---|---|
| own: | 1 4.7 cm anti-tank gun (t) with machine gun (t) |
| | 1 3.7 cm anti-tank gun (f) in revolving turret with machine gun (f) |
| | 2 heavy machine guns MG34 |
| | 2 light machine guns MG34 |
| | 2 defensive flame-throwers 42 |
| | 1 backpack radio set (d) |
| | 1 radio-telephone (C) |
| naval company: | 2 machine guns MG34 |
| VGAD (in case of combat): | 1 machine gun (Dutch) |

**4.)** **Military objectives**: The crew of the resistance nest will defend itself, to the last man, against enemy troops attacking from sea, air or land.

Specifically, any attempt by the enemy to land at North Beach on the northern flank of the harbour must be prevented.

Defence actions must focus on the entrance to the harbour.

The reserves are to be organised into assault detachments and held at the ready in this company sector.

**5.)** **Operations**: Enemy troops landing between North Mole and Salerie Corner are to be engaged using the 3.7 cm anti-tank gun (t) with machine gun, and the two heavy machine guns MG34. Enemy forces attacking on the coast road are to be repulsed using the 4.7 cm anti-tank gun with machine gun, as well as the two defensive flame-throwers 42.

The coast road is to be secured against mechanised attack by the 4.7 cm anti-tank gun (t). Any tanks that have broken through are to be destroyed by anti-tank weapons. In the event of a successful enemy breakthrough, the company commander will order the deployment of assault detachments against the advancing enemy. The defending forces will be supported by two 3.7 cm anti-aircraft guns and a 2 cm anti-aircraft gun at the harbour, as well as by a 2 cm anti-aircraft gun and a 7.5 mm machine gun at Salerie Corner.

Widerstandsnest  H A F E N R E S E R V E
(zugl. Kp.Gef.Stand)
_____

1.) <u>Einsatz</u> am Hafenvorplatz, mit Stoßtrupps im gesamten
Kompanieabschnitt.

2.) <u>Stärke:</u>
    eigene:             2 Offz. 7 Uffz. 25 Mannschaften
    Marine (nur bei
      Alarmstufe II):(2)Offz. 3 Uffz.  9 Mannschaften
    Im Kampffall treten hinzu die verfügbaren Kräfte des
    VGAD, des Hako und der im Hafen liegenden Boote.

  3.) <u>Waffen:</u>
    eigene:          1  4,7 cm Pak (t) mit M.G.(t)
                  1  3,7 cm Pak (f) im Drehturm mit
                                 M.G.(f)
                  2  s.M.G.34
                  2  le.M.G.34
                  2  Abwehrflammenwerfer 42
                  1  Tornisterfunkgerät (d)
                  1  Sprechfunkgerät (c)
    Marine-Komp.:    2  M.G.34
    VGAD (im Kampf=
        fall):      1  M.G. (holl.)

4.) <u>Kampfauftrag:</u> Die Besatzung verteidigt sich gegen von
See, Land und aus der Luft angreifenden Feind in ihrem
Widerstandsnest bis zum letzten Mann.

Im besonderen ist das Anlanden von Feind in der Nord=
flanke des Hafens North-Beach zu verhindern.

Verteidigungsschwerpunkt: Hafeneingang.

<u>Die Reserven sind stoßtruppartig</u> zu gliedern und zum
Einsatz im Kompanieabschnitt bereitzuhalten.

5.) <u>Kampfführung:</u> Anlandender Feind zwischen Nordmole und
Peterseck ist mit 1  3,7 cm Pak (t) mit M.G. und
2 s.M.G.34 zu bekämpfen. Auf der Uferstraße angreifen=
der Feind wird durch die 4,7 cm Pak mit M.G., sowie
durch 2 Abwehrflammenwerfer 42 bekämpft.

Die Uferstraße ist durch die 4,7 cm Pak (t) gegen an=
greifende Panzer gesichert. Durchgebrochene Panzer sind
mit Panzernahkampfmitteln zu vernichten. Bei gelungenem
Feindeinbruch werden auf Befehl des Komp.-Chefs Stoß=
trupps gegen den eingedrungenen Feind angesetzt.
Unterstützt wird der Abwehrkampf durch 2  3,7 cm Flak
und 1  2 cm Flak im Hafen, sowie durch 1  2 cm Flak und
1  7,5 mm M.G. auf Peterseck.

# SALERIE CORNER Resistance Nest

**1.)** __Deployment__ between Old Harbour and Gemäuer.

**2.)** __Contingent__:

navy:                                                    1 petty officer            10 ratings

fully manned only in the event of combat, at other times staffed by a skeleton crew of 1 petty officer and 3 ratings

**3.)** __Weapons__:

own:                              2 2 cm anti-aircraft guns
2 7.5 mm machine guns
1 machine gun MG311 (f) in revolving turret

**4.)** __Military objectives__: The crew of the resistance nest will defend itself, to the last man, against enemy troops attacking from sea, air or land.

Specifically, enemy troops landing between __North Mole and Gemäuer__ are to be resisted using all available weapons.

Defence actions must focus on harbour entrances and possible landing places between North Mole and Salerie Corner as well as in the vicinity of the water cooling pond.

**5.)** __Operations__: Enemy troops landing at the entrances between North Mole and Salerie Corner, as well as enemy troops landing in the vicinity of the water cooling pond, are to be engaged using two 2 cm anti-aircraft guns and the machine gun MG311 (f) in revolving turret. In addition, depending on the situation, the two 7.5 mm machine guns will be used. Tanks attacking on the coast road will be repelled using close-combat anti-tank weapons.

The defence will be supported on the one hand by a 3.7 cm gun in tank turret with machine gun and a machine gun MG34, both positioned at the Old Harbour resistance nest, and on the other hand by a 4.7 cm anti-tank gun and a light machine gun in a Tobruk pit at the Gemäuer resistance nest.

# Widerstandsnest  P E T E R S E C K

1.) <u>Einsatz</u> zwischen Hafenreserve und Gemäuer.

2.) <u>Stärke:</u>

    Marine                       1 Uffz.    1o Mannschaften

              Nur im Kampffall voll besetzt, sonst
              Sicherheitsbesatzung 1 Uffz. 3 Mannsch.

3.) <u>Waffen:</u>

    Marine:        2  2 cm Flak
                 2  M.G. 7,5 mm
                 1  M.G.311 (f) im Drehturm.

4.) <u>Kampfauftrag:</u> Die Besatzung verteidigt sich gegen von See, Land und aus der Luft angreifenden Feind in ihrem Widerstandsnest bis zum letzten Mann.

Im besonderen ist die Anlandung von Feindkräften zwi= schen <u>Nordmole und Gemäuer</u> mit allen Waffen zu bekämpfen.

Verteidigungsschwerpunkt: Einfahrten und Landestellen zwischen Nordmole und Peterseck und beim O.T.-Becken.

5.) <u>Kampfführung:</u> Gegen anlandenden Feind in den Einfahrten zwischen Nordmole und Peterseck, sowie gegen anlandenden Feind beim O.T.-Becken werden 2  2 cm Flak und 1 M.G.311 (f) im Drehturm eingesetzt, außerdem  je nach Lage die beiden M.G. 7,5 mm. Auf der Straße angreifende Panzer werden durch Panzernahkampfmittel bekämpft.

Unterstützt wird der Abwehrkampf einerseits durch eine 3,7 cm Kw.K. mit M.G. und durch 1 M.G.34 von Wider= standsnest Hafenreserve, andererseits durch eine 4,7 cm Pak und 1 le.M.G. im Tobrukstand von Widerstandsnest Gemäuer.

**Top**: View of Castle Cornet from Les Terres Point

**Middle**: A 3.7 cm anti-aircraft gun position at Les Terres Point

CASTLE CORNET VON LES TERRES POINT GESEHEN

3,7 cm FLAKSTELLUNG AUF LES TERRES POINT

Camouflaged casemate gun position at Soldiers Bay

GETARNTE K.-K.- STELLUNG IN DER SOLDATEN - BUCHT

Radar equipment at Fort George

FUNKMESS·GERÄTE IM FORT GEORGE

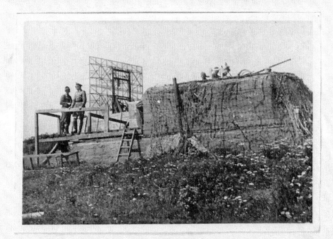

**Middle**: Outpost patrol boat sunk in the harbour during an air-raid

**Bottom**: Heavily armed convoy escorts

FLIEGER VERSENKTEN EIN V.P.-BOOT IM HAFEN

ZUM GELEITSCHUTZ EINGESETZTE ARTILLERIETRÄGER

**1**:   Albert Pier at low tide
**2**:   St Peter Port
**3**:   Model Yacht Pond – Victoria Tower in the background
**4**:   North Mole with explosive charges in place

ALBERT PIER BEI NIEDRIGWASSER

ST. PETER PORT

MODELLHAFEN    IM HINTERGRUND DER VICTORIA TURM

NORDMOLE ZUR SPRENGUNG VORBEREITET

**Top**: Boom defence between North and South Mole
**Bottom**: Havelet Bay

HAFENSPERRE ZWISCHEN NORD- UND
SÜDMOLE

HAVELET BAY

**Top**:    St James's Church and Elizabeth College
**Middle**: Looking north-east from Victoria Tower
**Bottom**: Looking north from Victoria Tower

ST. JAMES KIRCHE UND ELIZABETH COLLEGE

VOM VICTORIA-TURM NACH N.O.

VOM VICTORIA-TURM NACH NORD

**Top**: View of Castle Cornet from Havelet Bay
**Middle**: Sark and Herm in the background
**Bottom**: View of Castle Cornet from Belvedere House

CASTLECORNET VON DER HAVELET BAY

IM HINTERGRUND SARK UND HERM

CASTLE CORNET VOM BELVEDERE HAUS

**Top**: Access to Castle Cornet
**Middle**: Bastion on the north side
**Bottom**: Castle Pier

ZUGANG ZU CASTLE CORNET

BASTION AN DER NORDSEITE

CASTLE PIER

**Top**: Entrance to the barracks at Castle Cornet

**Middle**: Anti-aircraft gun emplacement and bastions on the castle walls

**Bottom**: Old lifts for transporting shells

KASERNENEINGANG IM CASTLE CORNET

FLAKSTAND UND BASTIONEN IM SCHLOSSINNEREN

ALTE GESCHOSSAUFZÜGE

**Top**: The oldest tower at Castle Cornet

**Left Middle**: Coat of arms above the main entrance

**Right Middle**: Sentry box built into the wall

**Bottom**: Old embrasures and ground platforms

DER ÄLTESTE TURM AUF CASTLE CORNET

WAPPEN ÜBER DEM HAUPTEINGANG

SCHILDERHAUS IN DER MAUER

ALTE SCHIESS-SCHARTEN UND GESCHÜTZBETTUNGEN

**Top**:     Inside view of a casemate gun bunker at Soldiers Bay
**Middle**:  Gun bunker on Castle Pier
**Bottom**:  7.5 cm gun used to protect the harbour entrance

IM K.K. BUNKER SOLDATENBUCHT

GESCHÜTZBUNKER AUF DEM CASTLE PIER

ZUM SCHUTZ DER HAFENEINFAHRT EINGESETZTE 7,5 KAN.

**Top**: White Hart Hotel converted into a bunker to accommodate the harbour reserve.
On the right: embrasure for 4.7 cm anti-tank gun
**Middle**: 4.7 cm anti-tank gun (t) securely built into the hotel wall

HOTEL ZUR UNTERBRINGUNG DER HAFENRESERVE ALS BUNKER UMGEBAUT · RECHTS SCHARTE F. 4.7 PAK

FESTUNGSMÄSSIG EINGEBAUTE 4,7 cm PAK (t)

Arrival at St Peter Port of a supply convoy from Jersey

EINLAUFEN EINES VON JERSEY KOMMENDEN VERSORGUNGS-GELEITES IN DEN HAFEN VON ST. PETER PORT

St. Peter Port

BRIT. KANALINSEL GUERNSEY

# BELLE GRÈVE BAY

from Gemäuer to Mont Crevelt

Belle Grève Bay
from Gemäuer to Mont Crevelt

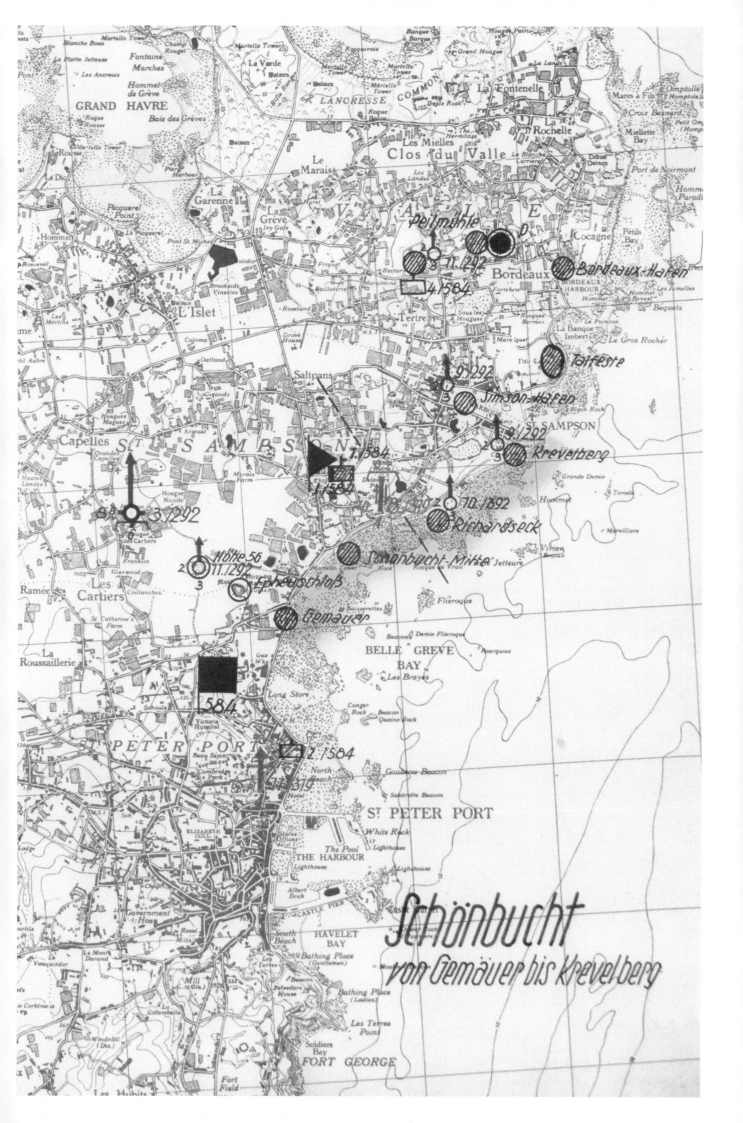

# St Sampson Harbour

## Legend

⊥ 10.5 cm gun in casemate
↑ 10.5 cm gun in field position
⊺ 7.5 cm gun in embrasure
⊥ 4.7 cm anti-tank gun in casemate
⊤ 3.7 cm and 5 cm anti-tank gun in field position
⊥ 3.7 cm tank gun
⅄ 3.7 cm tank gun and machine-gun
⊥ Armoured vehicle with machine-gun
∠ 8 cm and 5 cm mortars
◢ M19 (Maschinengranatwerfer, a fully-automatic 5 cm mortar)
◎ Tobruk pit
♪ Spigot mortar
✿ Multi-loopholed turret
ʄ Searchlights
⬭ Mines

**Firing zones**:
◁ 10.5 cm gun
◁ 4.7 cm anti-tank gun
⬭ Anti-aircraft gun
⬭ Mortar and M19
⬭ Machine-gun
⬭ Spigot mortar

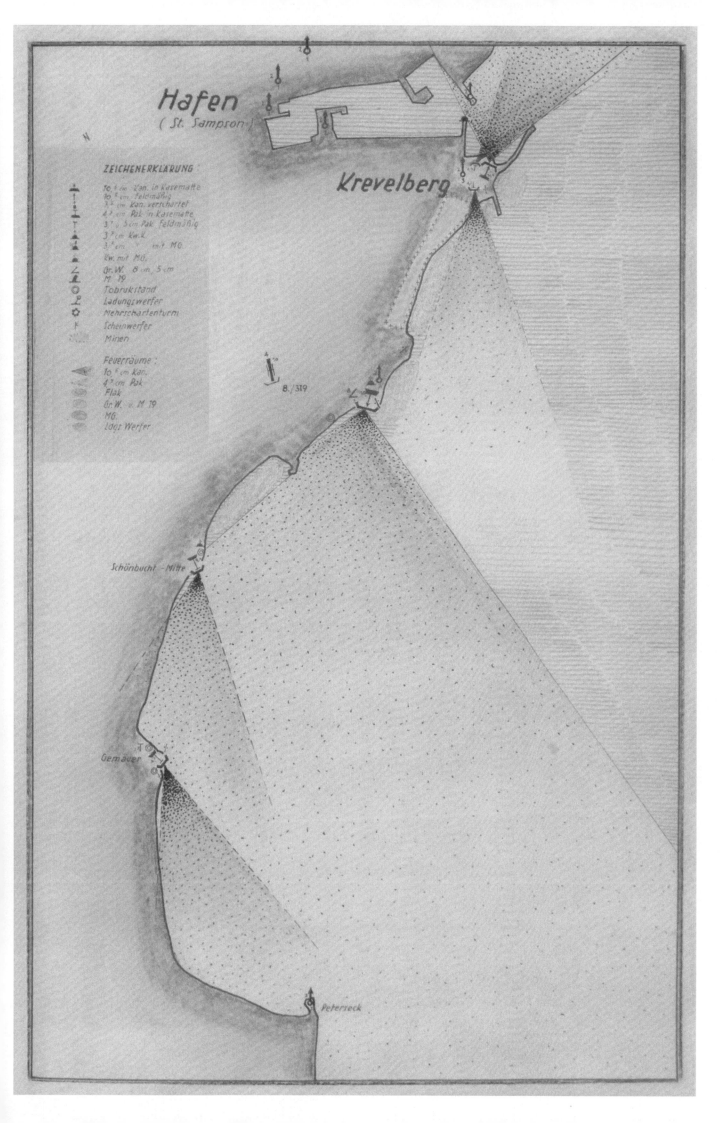

Hafen
( St. Sampson )

Krevelberg

ZEICHENERKLÄRUNG:

10,5 cm Kan. in Kasematte
10,5 cm feldmäßig
7,5 cm Kan. verschartet
4,7 cm Pak in Kasematte
3,7 u. 5 cm Pak feldmäßig
3,7 cm Kw.K.
3,7 cm " mit MG.
Kw. mit MG.
Gr.W. 8 cm 5 cm
M 19
Tobrukstand
Ladungswerfer
Mehrschartenturm
Scheinwerfer
Minen

Feuerräume:
10,5 cm Kan.
4,7 cm Pak
Flak
Gr.W. u. M 19
MG.
Ldgs.Werfer

8./319

Schönbucht - Mitte

Gemäuer

Peterseck

Schönbucht

Belle Grève Bay

Gemäuer infantry position

Infanterie-Stützpunkt „Gemäuer"

View of Belle Grève Bay from Cambridge Park

# GEMÄUER Resistance Nest

**1.)** <u>Deployment</u> between Salerie Corner and the middle of Belle Grève Bay.

**2.)** <u>Contingent</u>:

| | | |
|---|---|---|
| own | 1 NCO | 10 men |
| naval company | | |
| (only at stage two alert) | 2 petty officers | 8 ratings |

**3.)** <u>Weapons</u>:

own:
- 1 4.7 cm anti-tank gun (t) with machine gun (t)
- 1 3.7 cm anti-tank gun
- 2 machine guns MG34
- 2 defensive flame-throwers

naval company:
- 2 machine guns MG34

**4.)** <u>Military objectives</u>: The crew of the resistance nest will defend itself, to the last man, against enemy troops attacking from sea, air or land.

Specifically, enemy troops landing between <u>Salerie Corner and the middle of Belle Grève Bay</u> are to be resisted using all available weapons.

Defence actions must focus on entrance points and possible landing places in the vicinity of the water cooling pond and at Belle Grève Bay.

**5.)** <u>Operations</u>: Enemy troops landing at the entrance point near the water cooling pond are to be engaged using the 4.7 cm anti-tank gun with machine gun and a machine gun in a Tobruk pit. Enemy troops landing at Belle Grève Bay are to be engaged using a 3.7 cm anti-tank gun and a machine gun MG34 in a Tobruk pit. Both entrances are secured against mechanised attacks by armour-piercing weapons. Tanks attempting a breakthrough will be repulsed by anti-tank weapons. In addition, two defensive flame-throwers will be used for close defence.

The defence of the resistance nest will be supported by a 2 cm anti-tank gun, a machine gun MG311 (f) in revolving turret and a 7.5 mm machine gun, all positioned at Salerie Corner, as well as by the 4.7 cm anti-tank gun at the resistance nest at the middle of Belle Grève Bay.

Enemy forces that have penetrated as far as the Sports Ground, or airborne troops that have landed there, are to be repelled by sending assault detachments to engage them.

# Widerstandsnest " G E M Ä U E R "

1.) <u>Einsatz</u> zwischen Peterseck und Schönbucht-Mitte.

2.) <u>Stärke:</u>

    eigene                                    1 Uffz. 1o Mannschaften

    Marine-Komp.(nur bei
              Alarmstufe II)    2 Uffz.  8 Mannschaften

3.) <u>Waffen:</u>

    eigene:     1  Pak 4,7 cm(t) mit M.G.(t)
                1  Pak 3,7 cm
                2  M.G.34
                2  Abwehrflammenwerfer

    Marine Kp.:2  M.G.34

4.) <u>Kampfauftrag:</u> Die Besatzung verteidigt sich gegen von See, Land und aus der Luft angreifenden Feind in ihrem Widerstandsnest bis zum letzten Mann.

Im besonderen ist die Anlandung von Feindkräften zwi= schen <u>Peterseck und Schönbucht-Mitte</u> mit allen Waffen zu bekämpfen.

Verteidigungsschwerpunkt: Einfahrt und Landestellen beim O.T.-Becken und in der Schönbucht.

5.) <u>Kampfführung:</u> In der Einfahrt beim O.T.-Becken anlan= dender Gegner wird durch eine 4,7 cm Pak mit M.G. und durch 1 M.G. im Tobrukstand bekämpft. Anlandender Feind in der Schönbucht wird durch eine 3,7 cm Pak sowie durch 1 M.G.34 im Tobrukstand bekämpft. Beide Einfahr= ten sind durch panzerbrechende Waffen gegen anlandende Panzer gesichert. Außerdem werden durchbrechende Panzer mit Panzernahkampfmitteln bekämpft. Für Nahverteidigung werden zusätzlich 2 Abwehrflammenwerfer eingesetzt.

Unterstützt wird der Abwehrkampf des Widerstandsnestes durch 1  2 cm Flak, 1 M.G.311 (f) im Drehturm und 1 M.G. 7,5 mm von Widerstandsnest Peterseck, sowie durch die 4,7 cm Pak von Widerstandsnest Schönbucht-Mitte.

Eingedrungener oder aus der Luft gelandeter Feind im Sportgrund ist durch Entsenden von Stoßtrupps zu bekämp= fen.

# MID BELLE GRÈVE BAY Resistance Nest

**1.)**  **Deployment** at the left-hand boundary of the Harbour Company, between Gemäuer and Richmond Corner.

**2.)**  **Contingent**:

own                                        1 NCO                7 men

**3.)**  **Weapons**:

own:                    1 4.7 cm anti-tank gun (t) with machine gun (t)
                        1 machine gun MG311 (f) in revolving turret

**4.)**  **Military objectives**: The crew of the resistance nest will defend itself, to the last man, against enemy troops attacking from sea, air or land.

Specifically, enemy troops landing between Gemäuer and Richmond Corner are to be resisted using all available weapons.

Defence actions must focus on entrance points and possible landing places between this resistance nest and Gemäuer.

**5.)**  **Operations**: Enemy troops landing at Belle Grève Bay are to be engaged using the 4.7 cm anti-tank gun (t). Enemy troops landing at the entrance between the middle of Belle Grève Bay and Richmond Corner are to be engaged using the machine gun MG311 (f) in revolving turret. Belle Grève Bay is vulnerable to mechanised attacks; therefore the 4.7 cm anti-tank gun (t) will be used to secure it against tanks. Otherwise tanks will be repulsed by anti-tank weapons.

The defence of the resistance nest will be supported by the 3.7 cm anti-tank gun positioned at Gemäuer, as well as the 10.5 cm casemate gun and a 2 cm anti-aircraft gun positioned at Richmond Corner.

# Widerstandsnest  S C H Ö N B U C H T - M I T T E

1.) <u>Einsatz</u> an linker Grenze Hafen-Kompanie, zwischen
Gemäuer und Richardseck.

2.) <u>Stärke:</u>
    eigene:                     1 Uffz.  7 Mannschaften.

3.) <u>Waffen:</u>
    eigene:    1  4,7 cm Pak (t) mit M.G. (t)
                1  M.G. 311 (f) im Drehturm.

4.) <u>Kampfauftrag:</u> Die Besatzung verteidigt sich gegen von
See, Land und aus der Luft angreifenden Feind in ihrem
Widerstandsnest bis zum letzten Mann.

Im besonderen ist die Anlandung von Feindkräften zwischen
<u>Gemäuer und Richardseck</u> mit allen Waffen zu bekämpfen.

Verteidigungsschwerpunkt: Einfahrt und Landestellen
zwischen dem Widerstandsnest und Gemäuer.

5.) <u>Kampfführung:</u>  In der Schönbucht anlandender Gegner
wird mit der 4,7 cm Pak (t), in der Einfahrt zwischen
Schönbucht-Mitte und Richardseck anlandender Feind mit
M.G.311 (f) im Drehturm bekämpft. Die panzergefährdete
Schönbucht wird gegen Panzer durch die 4,7 cm Pak (t)
gesichert, im übrigen werden Panzer mit Panzernahkampf=
mitteln bekämpft.

Unterstützt wird der Kampf des Widerstandsnestes durch
die 3,7 cm Pak auf Gemäuer und durch die 1o,5 cm Kas.Kan.
sowie eine 2 cm Flak auf Richardseck.

# RICHMOND CORNER Resistance Nest

**1.)** **Deployment** at the right-hand boundary of the North Company between the middle of Belle Grève Bay and Mont Crevelt.

**2.)** **Contingent**:

| | | |
|---|---|---|
| own: | 2 NCOs | 15 men |
| artillery: | 1 NCO | 3 men |
| anti-aircraft: | 4 NCOs | 18 men |

**3.)** **Weapons**:

| | |
|---|---|
| own: | 1 10.5 cm casemate gun (f) (in fortified emplacement) |
| | 1 machine gun MG311 (f) in tank turret |
| | 1 8 cm mortar |
| | 3 light machine guns MG34 (usable as heavy machine guns when on bunker mount) |
| | 1 loophole searchlight |
| | 5 defensive flame-throwers 42 |
| artillery: | 1 60 cm searchlight |
| anti-aircraft: | 3 2 cm anti-aircraft guns |
| | 1 machine gun MG34 |
| | 1 60 cm searchlight |

**4.)** **Military objectives**: The garrison of the resistance nest will defend itself, to the last man, against enemy troops attacking from sea, air or land.

Specifically, enemy troops landing between <u>the middle of Belle Grève Bay and Mont Crevelt</u> are to be resisted using all available weapons.

Defence actions must focus on entrance points and possible landing places between this resistance nest and the middle of Belle Grève Bay.

**5.)** **Operations**: Enemy troops landing at the right-hand flank between the middle of Belle Grève Bay and Richmond Corner are to be engaged using the casemate gun and two heavy machine guns. Landing attempts between Richmond Corner and Mont Crevelt will be repelled using a heavy machine gun. Dead spaces will be covered by the 8 cm mortar. Supporting fire for the resistance nest will be provided by a 4.7 cm anti-tank gun with machine gun and a heavy machine gun positioned at Mont Crevelt, plus two 8 cm mortars positioned at Galgenberg and Peilmühle respectively.

In order to guard against low-level attacks by enemy aircraft and the landing of airborne troops, an anti-aircraft platoon, equipped with three 2 cm guns, is to be deployed at the resistance nest. Unless these guns are engaged in firing at enemy aircraft, they will be used for the defence at ground level.

A tank destroyer squad will be assigned for anti-tank defence at the rear of the resistance nest. Enemy forces landing outside the area of this resistance nest will be engaged immediately by an assault detachment.

# Widerstandsnest R I C H A R D S E C K

1.) **Einsatz** an rechter Grenze Nord-Kompanie zwischen Schönbucht-Mitte und Krevelberg.

2.) **Stärke:**

|  |  |  |
|---|---|---|
| eigene: | 2 Uffz. | 15 Mannschaften |
| Artl.: | 1 Uffz. | 3 Mannschaften |
| Flak: | 4 Uffz. | 18 Mannschaften |

3.) **Waffen:**

| eigene: | 1 | Kas.Kan. 1o,5 cm (f) (festungsmäßig) |
|---|---|---|
|  | 1 | M.G.311 (f) i/Panzerkuppel |
|  | 1 | 8 cm Gr.Werf. |
|  | 3 | le.M.G.34 (auf Bunkerlafette als s.M.G.) |
|  | 1 | Schartenscheinwerfer |
|  | 5 | Abwehr-Flammenwerfer 42 |
| Artillerie: | 1 | Scheinwerfer 6o cm |
| Flak: | 3 | 2 cm Flak |
|  | 1 | M.G.34 |
|  | 1 | Scheinwerfer 6o cm |

4.) **Kampfauftrag:** Die Besatzung verteidigt sich gegen von See, Land und aus der Luft angreifenden Feind in ihrem Widerstandsnest bis zum letzten Mann.

Im besonderen ist die Anlandung von Feindkräften zwi= schen Schönbucht-Mitte und Krevelberg mit allen Waffen zu bekämpfen.
Verteidigungsschwerpunkt: Einfahrten und Landestellen zwischen dem Widerstandsnest und Schönbucht-Mitte.

5.) **Kampfführung:** In der rechten Flanke zwischen Schönbucht-Mitte und Richardseck anlandender Gegner wird von einer Kas.Kan. und zwei s.M.G. bekämpft. Eine zwischen Richardseck und Krevelberg versuchte Landung wird durch ein s.M.G. bekämpft. Tote Räume werden durch einen 8 cm Gr.Werf. ausgeschaltet. Das Widerstandsnest wird in seinem Feuerkampf unterstützt durch eine 4,7 cm Pak mit M.G. und ein s.M.G. von Krevelberg, je einen 8 cm Gr.Werf. von Galgenberg und Peilmühle.

Zur Sicherung gegen Tieffliegerangriffe und Landung aus der Luft ist auf dem Widerstandsnest ein Zug Flak mit 3 2 cm Geschützen eingesetzt. Sind diese Geschütze nicht durch Flugzeugbeschuß gebunden, greifen sie mit in den Erdkampf ein.

Zur Panzerabwehr nach rückwärts ist ein Panzervernich= tungstrupp eingeteilt. Gelandeter Gegner außerhalb des Widerstandsnestes wird sofort durch eine Stoßgruppe bekämpft.

# MONT CREVELT Resistance Nest

**1.)** **Deployment** at the entrance to St Sampson Harbour between Richmond Corner and Vale Castle.

**2.)** **Contingent**:

| | | | |
|---|---|---|---|
| own: | | 2 NCOs | 12 men |
| anti-aircraft artillery: | 1 officer | 3 NCOs | 17 men |
| navy: | | | 3 ratings |

**3.)** **Weapons**:

own:
- 1 10.5 cm casemate gun (in fortified emplacement)
- 2 4.7 cm anti-tank guns (t) with machine gun (in reinforced field mountings)
- 1 3.7 cm gun with machine gun MG311 (f) in tank turret
- 1 heavy machine gun MG34
- 3 defensive flame-throwers 42

anti-aircraft:
- 3 2 cm anti-aircraft guns
- 1 machine gun MG34
- 1 60 cm searchlight

of the above, one gun and the searchlight are deployed at the North Mole and Middle Pier at St Sampson Harbour.

**4.)** **Military objectives**: The crew of the resistance nest is to defend itself, to the last man, against enemy troops attacking from sea, air or land.

Specifically, enemy troops landing at St Sampson Harbour and between Richmond Corner and Vale Castle, are to be resisted using all available weapons.

Defence actions must focus on the entrance to St Sampson Harbour.

**5.)** **Operations**: Enemy troops landing between Richmond Corner and Mont Crevelt are to be engaged using a 4.7 cm anti-tank gun with machine gun and a heavy machine gun. Enemy forces that have landed will be delayed by a minefield laid with assorted mines. Enemy craft approaching the entrance to St Sampson Harbour will be repulsed using the 10.5 cm casemate gun, a 4.7 cm anti-tank gun with machine gun, and the 3.7 cm anti-tank gun in tank turret.

There is a mine barrier, laid by the navy, approximately 400 metres in front of the harbour entrance. From the fortified ignition control panel, situated roughly 50 metres to the right of the casemate gun bunker, these mines can be detonated individually as required.

In addition, the following weapons will cover St Sampson Harbour: two light machine guns and one machine gun MG311 (f) in tank turret, positioned at the St Sampson Harbour resistance nest; a 10.5 cm casemate gun and one heavy machine gun positioned at Vale Castle.

Enemy landing attempts will be further thwarted by one 8 cm mortar located at Galgenberg, one 8 cm mortar at Peilmühle, one 8 cm mortar at Richmond Corner, and three 5 cm mortars at Vale Castle.

Enemy troops who have landed and established themselves in the houses, the majority of which are unoccupied, will be engaged by a counter-attack detachment from this resistance nest. A tank destroyer squad will repel mechanised attacks from the rear of the resistance nest.

In order to guard against low-level attacks by enemy aircraft, an anti-aircraft platoon, equipped with three 2 cm guns, is to be deployed at the resistance nest.

# Widerstandsnest K R E V E L B E R G

1.) <u>Einsatz</u> an Hafeneinfahrt Sampson-Hafen zwischen Richards=eck und Talfeste.

2.) <u>Stärke:</u>

| | eigene: | | 2 Uffz. | 12 Mannschaften |
|---|---|---|---|---|
| | Flak: | 1 Offz. 3 Uffz. | 17 Mannschaften |
| | Marine: | | | 3 Mannschaften |

3.) <u>Waffen:</u>
   eigene:
   1 Kas.Kan. 1o,5 cm (festungsmäßig)
   2 Pak 4,7 cm (t) mit M.G. (feldm.verstärkt)
   1 Kwk.3,7 cm mit M.G.311 (f) in Panzerkuppel
   1 s.M.G.34
   3 Abwehrflammenwerfer 42

   Flak:
   3 2cm Flak
   1 M.G.34
   1 Scheinwerfer 6o cm,

   davon sind 1 Geschütz und Scheinwerfer auf
   Nordmole und Mittelpier Sampsonhafen eingesetzt

4.) <u>Kampfauftrag:</u> Die Besatzung verteidigt sich gegen von See, Land und aus der Luft angreifenden Feind in ihrem Wider=standsnest bis zum letzten Mann.

   Im besonderen ist die Anlandung von Feindkräften im Hafen St.Sampson und zwischen <u>Richardseck und Talfeste</u> mit allen Waffen zu bekämpfen.

   Verteidigungsschwerpunkt: Einfahrt St.Sampson-Hafen.

5.) <u>Kampfführung:</u> Ein zwischen Richardseck und Krevelberg anlandender Gegner wird bekämpft von einer 4,7 cm Pak mit M.G. und einem s.M.G. Angelandeter Feind wird durch ein gemischtes Minenfeld aufgehalten. Sich der Hafeneinfahrt St.Sampson nähernde Feindboote werden bekämpft durch eine Kas.Kan. 1o,5 cm, eine 4,7 cm Pak mit M.G., eine 3,7 cm Pak in Panzerkuppel.

   In dem Raum etwa 400 m vor der Hafeneinfahrt liegt eine von der Marine verlegte Minensperre, die bei Bedarf aus dem etwa 50 m rechts vom Kas.Kan.Bunker befindlichen festungsmäßigen Minenzündtisch einzeln gezündet werden kann.

   Folgende Waffen wirken außerdem in den Hafen St.Sampson: 2 le.M.G. und 1 M.G.311 (f) in Panzerkuppel aus dem Wider=standsnest Simson-Hafen, eine 1o,5 cm Kas.Kan. und 1 s.M.G. von Talfeste (Vale Castle).

   Angelandeter Gegner wird außerdem bekämpft durch 1 8 cm Granatwerfer von Galgenberg, 1 8 cm Granatwerfer von Peil=mühle, 1 8 cm Granatwerfer von Richardseck und 3 5 cm Granatwerfer von Talfeste.

   An Land gekommener Feind, der die Möglichkeit hat, sich in den zum größten Teil leerstehenden Häusern festzusetzen, wird durch eine Gegenstoßgruppe aus dem Widerstandsnest angegriffen. Von rückwärts angreifende Panzer werden durch einen Panzervernichtungstrupp bekämpft.

   Zur Sicherung gegen Tieffliegerangriffe ist ein Zug Flak mit 3 2 cm Geschützen auf dem Widerstandsnest eingesetzt.

1: View from Infantry Regiment 584's command post to Belle Grève Bay and the northern part of the island
2: Belle Grève Bay with the resistance nest at the middle of Belle Grève Bay. On the hill, the field emplacement of 8th Battery, Artillery Regiment 319
3: Gemäuer strong point
4: View of the town and Salerie Point from Gemäuer strong point

BLICK VOM GEFECHTSSTAND I.R. 584 AUF SCHÖNBUCHT U. NORDTEIL
DER INSEL.

DIE SCHÖNBUCHT MIT STÜTZPUNKT SCHÖNBUCHT MITTE. AUF DEM BERG FEUERSTELLUNG
DER 8./A.R. 319

STÜTZPUNKT „GEMÄUER".

BLICK VOM STÜTZPUNKT GEMÄUER AUF STADT UND PETERSECK

**Top**: Embrasure for a 4.7 cm anti-tank gun (t) at the middle of Belle Grève Bay

**Middle**: Looking north from the base at Richmond Corner

**Bottom**: Emplaced defensive flame-thrower 42

SCHARTE FÜR 4,7 PAK (t) AUF SCHÖNBUCHT MITTE

STÜTZPUNKT RICHARDSECK · BLICK NACH NORDEN

EINGEBAUTER ABWEHRFLAMMENWERFER 42

**Top**:     Entrance to Richmond Corner with tank turret and machine gun MG311 (f)
**Middle**:  Field of fire of the casemate guns covering Belle Grève Bay and the town
**Bottom**:  In the background, Mont Crevelt and the entrance to St Sampson Harbour

EINGANG ZU RICHARDSECK MIT PANZERKUPPEL U. M.G.311 (f)

SCHUSSFELD DER K.K. AUF SCHÖNBUCHT U. STADT

IM HINTERGRUND „KREWELBERG" U. EINFAHRT SAMPSON HAFEN

STÜTZPUNKT KREWELBERG MIT SAMPSON-HAFEN EINFAHRT. IM HINTERGRUND DIE INSEL HERM

BLICK VOM KREWELBERG AUF DIE SCHÖNBUCHT U. ST. PETER PORT. IM HINTERGRUND HERM U. SARK

**Top**: The base at Mont Crevelt, with the entrance to St Sampson Harbour. The island of Herm in the background
**Bottom**: View of Belle Grève Bay and St Peter Port from Mont Crevelt. Herm and Sark in the background

To subscribe and support this part work edition please
visit our website where you will find details of how
to subscribe and how the edition will unfold.

**www.clearvuepublishing.com**

**or email us at: info@clearvuepublishing.com**

Alternatively you can write to us:
The Clear Vue Publishing Partnership Ltd,
La Battue, St Peter Port,
Guernsey,
GY1 1UP

Our film on the history of the fortifications,
**Hitler's Island Madness**,
features images from the Festung
and is available in shops
and on our website
as is our three part oral history
**The Occupation of the Channel Islands**

Visitors to the Channel Islands can
Find out more about the fortifications at:

The Channel Islands Occupation Society [Guernsey]
www.occupied.guernsey.net

The Channel Islands Occupation Society [Jersey]
www.ciosjersey.org.net

Festung Guernsey
www.festungguernsey.supanet.com

First published 2007
in a limited edition of 135 copies

This paperback partwork edition
Published 2012

The Clear Vue Publishing Partnership Ltd
La Battue, Candie Road, St Peter Port
Guernsey, Channel Islands
GY1 1UP

Festung Guernsey reproduced by kind permission of: The Royal Court 2007

**www.clearvuepublishing.com**